Contents

Some words are shown in bold, **like this.** You can find out what they mean by looking in the glossary.

What Is Soil?

Soil is one of Earth's many natural resources. We sometimes call it earth, dirt, or mud. It is the top layer of the land. It is the natural material that grass, trees, crops, and other plants grow in. Soil may be just a few inches deep or it may be many feet deep.

What is soil made from?

Soil is made from a mixture of sand, small gritty pieces of rock, water, and dead plant material called **humus**. The amounts of the different materials in soil vary from place to place. These materials stick to each other and act like a sponge, soaking up water. The water in soil contains natural chemicals. Some of these are taken up by plants and used as food. These substances are called nutrients. People need some of these substances, too. We get them by eating plants or animals that eat plants.

Soil is important for growing plants, including the vegetables, fruit, and cereals that we eat.

4

What is under the surface?

Soil is not the same all the way through. From the top to the bottom, it is divided into layers. These are the **litter zone,** topsoil, and subsoil.

The litter zone is made of leaves and other parts of plants that have fallen on the ground. They slowly break down and form humus. Below that, there is the topsoil. At the top, it is dark, crumbly, and spongy—a mixture of grit and humus. There is air and water in the spaces between the grit and humus. Deeper down, topsoil contains less humus. The weight of the soil above presses the **particles** closer together, so there is less air, too. Below the topsoil, there is the subsoil. The subsoil is even more pressed down and contains very little humus or nutrients.

litter zone

topsoil

subsoil

rock

Soil is divided into layers. The topsoil contains the most humus and nutrients.

What else does soil contain?

Soil contains living things, along with sand, grit, **humus,** water, and air. Most of the living things are so small that you cannot see them. The smallest are bacteria, **organisms** that can cause disease, and microscopic algae, or simple green plants without stems, roots, or leaves. There are also tiny animals—mostly worms called nematodes and other organisms called tardigrades. Tardigrades are called water bears because they look like tiny bears. They are actually eight-legged creatures less than 0.04 inch (1 millimeter) long. They live in the water between the soil **particles.** Bigger earthworms crawl through the topsoil. Slugs, snails, insects, and other creatures live in the cracks in the soil's surface.

Did you know?

There are more bacteria in a handful of soil than there are people on Earth!

Soil is full of living things.

6

Where Does Soil Come From?

The raw materials for soil are being made all the time. Big rocks are continually being worn down into smaller particles. This happens because earthquakes, **volcanoes,** and landslides break pieces off of large rocks. If water gets into cracks in rock and freezes, it expands and pieces of rock break off. All of these natural **processes** wear rock down and produce the tiny particles that help to make soil. Plants die or their leaves fall off. **Microbes** and **fungi** break down the dead plant material to produce humus. Wind blows the rock particles and dead plant material around until they settle somewhere. Plants grow into them, binding them together with their roots and forming new soil.

Many rocks in rivers are round because water moves them around and bangs them into each other, wearing down the sharp corners. The small pieces that are broken off help to form new soil.

7

How can floods improve soil?

We usually think of floods as bad things, but people in parts of the ancient world sometimes relied on floods to grow crops to eat. When rivers flood, they spread mud on the land. The mud is rich in **organic** material. As this rots, it breaks down into simple substances that plants can take up through their roots and use as food.

In some places, farmers still rely on mud from floods to enrich their soil. Frequent floods in Bangladesh cause great hardship to the people who live there, but the river mud they spread over the land also improves the soil. This enables farmers in Bangladesh to grow up to three crops in one year.

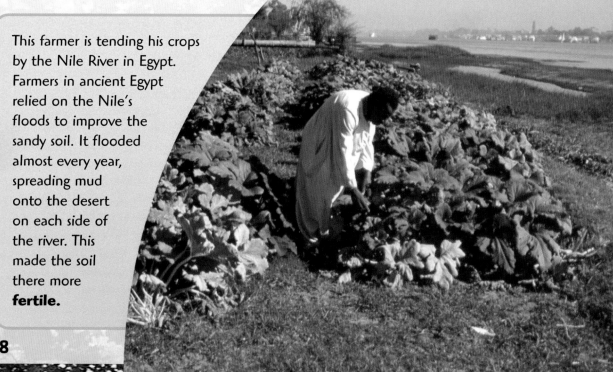

This farmer is tending his crops by the Nile River in Egypt. Farmers in ancient Egypt relied on the Nile's floods to improve the sandy soil. It flooded almost every year, spreading mud onto the desert on each side of the river. This made the soil there more **fertile.**

8

What Are the Different Kinds of Soil?

Soil is not the same everywhere. It depends on the raw materials that nature provides and the amounts of these raw materials differ from place to place. The main types of soil are: sandy, clay, chalky, peaty, and loam.

Sandy soil

Sandy soil has a lot of sand in it. The extra sand lets water drain through easily, so it dries out quickly after a rainstorm.

Clay soil

Clay soil has a lot of very small rock **particles.** They pack together so tightly that they make the soil very heavy and sticky. If clay dries out, it becomes as hard as stone.

Every country is covered with a patchwork of different kinds of soils. In this map of Australia, each color represents a different kind of soil.

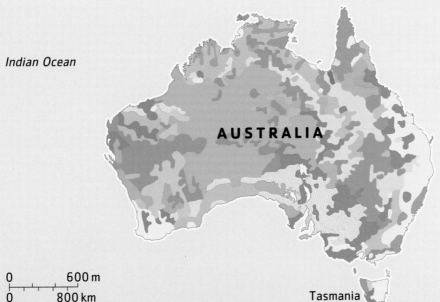

Indian Ocean

AUSTRALIA

Pacific Ocean

Tasmania

0   600 m
0   800 km

9

Chalky soil

Chalky soil has pieces of **limestone** in it. It is a bit like sandy soil and a bit like clay soil. Like sandy soil, water drains through chalky soil very quickly. Like clay soil, it can sometimes be sticky and hard to dig into.

Peaty soil

Peaty soil contains a lot of dead plant material. It is found in places where the soil is waterlogged, such as bogs. Peaty soil soaks up a lot of water like a sponge. Peat bogs, or wet, spongy grounds, grow only about 0.04 inch (1 millimeter) every year. Digging out peat can destroy a peat bog and its special **habitat** for plants and animals. Taking peat from nature is strictly controlled by law to protect peat bogs.

Peat dug out of the ground like this can be dried and burned like coal.

Loam

The best soil for growing most plants is called loam. Loam is a spongy mixture of sand, grit, clay, and **humus**. It does not dry out quickly or become waterlogged. It holds the right amount of water, air, and nutrients for most plants to grow well.

How Do Plants and Animals Use Soil?

Soil is necessary for most of the plant and animal life on land. Plants take up water and nutrients from the soil and use them for growth. Humans and animals take in these nutrients themselves when they eat plants.

Most people also eat meat. The supply of meat depends on plants because animals such as cows and pigs eat plants. In other words, our food chain begins with plants, and plants depend on soil.

Many animals make their homes in soil—from tiny creatures such as worms, spiders, and ants, to larger creatures such as moles, rats, rabbits, foxes, and badgers. Some ants make amazingly complicated habitats in the ground, where thousands of ants live together.

Rabbits live in burrows, or holes, that go deep into the ground.

Ants like these live in large colonies in the ground.

11

How do birds use soil?

Most birds build nests from the leaves, twigs, feathers, and other materials they find in the soil's **litter zone**. Some birds, including swallows, use the soil itself to build their homes. They pick up muddy soil in their beaks and make mud nests. The nests are often attached to trees, cliffs, and buildings. Many birds swallow grit and **gravel** and use them to grind up their food. Some parrots in South America use clay soil in a very strange way. They eat it! They need it because the clay absorbs poison from the fruits and seeds the birds eat. Some medicines for people with stomach problems contain the same type of clay, which is called kaolin.

Some birds build nests from mud.

12

Why is soil important to reptiles?

Scaly-skinned animals, such as snakes, turtles, lizards, and crocodiles, are called reptiles. Most reptiles lay eggs and some build nests for their eggs. The Nile crocodile digs a hole in the ground, lays her eggs in it, and then fills in the hole with soil. Other kinds of crocodiles and alligators pile up soil, mud, and plants and lay their eggs in the middle.

Why are there worms in soil?

Worms not only live in soil, they also find all their food in it. As earthworms move through the ground, they take in soil. They keep anything they can eat and let the rest pass out of their bodies. Worms are good for soil because they carry **organic** material down from the surface and mix up the soil.

Did you know?

Healthy soil is full of worms of all sizes. Earthworms take in their own weight of soil every day.

What Do We Use Soil For?

People have used soil for thousands of years. They piled it up to make banks to mark out their land and defend their homes. They used it for growing plants to eat and to feed their animals. They also used clay soil to make bricks, pots, and bowls. And they mixed colored soils with water to make paints to decorate their bodies. Soil and the materials it contains are still used to do many of these things today.

People have been making pottery from clay for about 10,000 years. This vase was made by the Romans in the 2nd century AD.

What is Offa's Dike?

Offa was a king in England more than 1,200 years ago. To mark his kingdom, he ordered a great bank of soil to be built. It stretched across the countryside for 167 miles (270 kilometers). It was up to 60 feet (18 meters) high. Another name for a bank of soil is a dike, so it was called Offa's Dike.

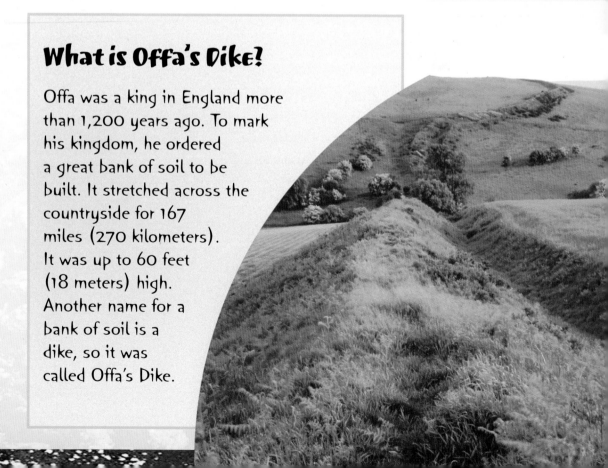

How is soil used in the home?

Some of the things we use every day, including plates and cups, are made from clay. There are different kinds of clay. The finest cups and plates are made from clay with the smallest **particles**. This is called china clay. It starts off as a soft, white powder. When it is mixed with water, it sticks together and can be molded into a shape. It is then heated to more than 1,800 °F (1,000 °C) so that some of it melts and changes to glass. This makes the clay hard and waterproof. Clay is also used in the production of rubber, paint, ink, paper, and some makeup.

Did you know?

The largest particles in china clay are only 0.0004 inch (0.01 millimeter) across.

15

How is soil used to make new land?

Parts of some coastlines are washed away by waves. But land does not have to be lost to the sea forever. The Dutch people have turned an enormous area of sea around the Netherlands' coast back into dry land. More than one sixth of the country was once under the sea.

The Dutch reclaimed each piece of land by building a bank of soil around it. Then the water was taken out, leaving dry land. At first, this was done by hand. Then, windmills were used to pump the water out. Now, the latest digging and pumping machines are used. Today, the low-lying parts of the Netherlands are protected by 1,800 miles (3,000 kilometers) of sand dunes and earth banks, or dikes.

Did you know?

Nearly two thirds of the people in the Netherlands live on land that is below sea level.

The Afsluitdijk (see page 17)

North Sea

THE NETHERLANDS

The Dutch have reclaimed a huge area of land from the sea. Today, low-lying areas are protected by a series of sand dunes and dikes.

Key
- land reclaimed from sea since 1200
- sand dunes
- dikes

0   25 m
25 km

BELGIUM

16

CASE STUDY:
The Afsluitdijk, The Netherlands

Afsluitdijk is a Dutch word meaning "enclosing dam." It is a bank of soil in the Netherlands that stretches for 19 miles (30 kilometers) across an **inlet** of the North Sea. It was completed in 1933 after 6 years of work by 5,000 people. Since then, much of the land behind the dam has been drained and reclaimed from the sea.

The dam was built from a material called till, a type of clay, held together with 18 million bundles of brushwood. Its slopes are covered with blocks of rock that prevent the clay from being washed away. Grass on top of the dam helps to bind it together and stop it from being **eroded** by wind and rain. It is so wide that a road and bicycle path have been built on top of it.

The Afsluitdijk is strong enough to hold back the North Sea.

Why Is Soil Important to Scientists?

Scientists can learn more about Earth, our history, and recent events by studying soil. The mixture of materials in soil gives clues about the natural **processes** that shape Earth. Soil may contain ash from an ancient **volcano, silt** from a flood long ago, or burnt **particles** from a great fire.

Soil on shoes or tires can prove where a person has been recently.

Did you know?

Soil can help to solve crimes. When you walk across a muddy field, some of the mud sticks to your shoes. It shows where you have been. After a serious crime, any soil found on people's clothes or car tires is collected and studied by scientists. The materials, seeds, and **organisms** it contains may prove where someone has walked. It may link them to a crime.

18

What does soil tell us about the past?

Digging down through soil is like traveling back in time. Scientists who try to understand the past by digging into the ground are called archaeologists. They scrape away the soil, layer by layer, and find parts of old buildings, everyday objects, and even skeletons from long ago. The deeper they dig, the older the things they find. Some things they look for are difficult to see and very easily destroyed. Dark stains in the soil may show that wooden posts there held up a building thousands of years ago. Blackened soil may show where a fire was made and food was cooked. If the soil is washed away or dug up, this important historical information could be lost forever.

Soil may contain clues to events long ago, such as brush fires.

How Does Soil Move Around?

Soil is moved great distances by the forces of nature. For example, dry soil can be blown into the air by the wind. The smallest **particles** take a long time to fall back to the ground. Sunlight passing through them may change colors and produce beautiful sunsets.

Rivers can move large amounts of soil hundreds of miles.

Soil can also be moved by landslides and water. If soil is washed into rivers, it quickly separates into its different parts. Sandy particles fall to the bottom and plant material floats. You can see this if you put some soil in a bottle of water, shake it up, and watch the bottle. The sand, grit, and plant material in soil can be carried a long way by rivers or sea currents before they reach land again. There, they may help to form new soil.

How do people move soil?

Soil is transported by trucks. When new homes are built, new soil is brought in to create the lawns and gardens. It is spread on the ground so that grass and other plants will grow well.

Trucks like this one can dump their loads of soil where it is needed.

20

CASE STUDY:
The Canterbury Plains, New Zealand

The Canterbury Plains form a vast, flat area of land on New Zealand's South Island. The plains are the result of about two million years of soil and gravel being moved by nature from one place to another.

Glaciers and rivers flowing down from the mountains of the Southern Alps carried gravel and fine **silt** down the mountains. Over time, the silt became so deep that it covered all the hills and valleys of the ground below. The rivers spread out across the flattened land, dropping more and more **sediment**. Today, the sediments that form the Canterbury Plains are more than 490 feet (150 meters) deep in some places.

The soil that covers New Zealand's Canterbury Plains was brought down from the Southern Alps mountain range by rivers.

21

How Can Soil Be Damaged?

Soil can be damaged in many ways. Some of the damage is caused by nature and some is caused by people.

A long period of time with no rain, or a drought, can dry out soil so much that wind can blow it away. Heavy rain can wash soil away, too. Losing soil by wind or water is called **erosion**. When people cut down trees or clear land for farming, the soil is no longer held together by the roots of plants. This makes soil erosion more likely to happen.

Many of the chemicals used in factories are harmful to soil. In the past, some factories poured waste chemicals into the ground. Sometimes chemicals were buried in the ground to get rid of them, but their containers leaked and the soil was **polluted.** Today, many plants and creatures that live in soil are killed by pollution. Some plants grow and take up the pollution through their roots. People and other animals who eat plants from polluted soil can become very sick.

Soil is likely to erode if there are no plants or grass in it.

22

How can watering the ground harm soil?

In places where the land is too dry to grow crops, extra water is sometimes supplied by wells or pipelines. Watering land for farming is called irrigation. Irrigation water contains a tiny amount of salt. Over a long time, the salt can build up in the soil. However, most plants and **organisms** cannot live in salty soil. Salt pollution is a serious problem in places such as Pakistan, where irrigation is necessary for farming.

Rain running through polluted soil can carry the pollution into nearby rivers.

How Do Farmers Take Care of Soil?

In nature, the nutrients and **organic** matter in soil are constantly **recycled**. They are used by plants and animals and then returned to the soil when plants die or as bodily waste from animals. On a farm, nutrients and organic matter are taken away every time crops are harvested. The nutrients have to be replaced. Farmers use **fertilizer** to help the soil produce good crops. However, some of the fertilizers and chemicals used on farms can be washed off the land by rain and into **reservoirs** that people's drinking water comes from.

Do organic farmers look after soil differently?

Organic farmers use as few chemicals as possible. They try to keep the soil as natural as possible so that safer and healthier food is produced.

Organic farmers let shrubs, low trees, and wild plants grow around their fields to reduce soil **erosion** and provide **habitats** for creatures that control pests. These vegetables have been mixed with marigolds.

24

How Does Soil Affect the Way We Live?

Soil affects our lives in many ways. It is at the heart of most of the world's food production. Buildings, roads, and bridges are built on it. Many of the materials our homes are built from depend on soil or materials that are grown in soil, such as wood.

How does soil affect buildings?

If soil dries out quickly, it can sink. If soil soaks up a lot of extra water quickly, the ground can move up or rise. If the soil sinks or moves up, it can damage buildings. It is important for builders to understand what the ground is like before they start building on it.

If the ground dries out or swells up with water, it can move enough to crack a building's walls.

CASE STUDY:
The Leaning Tower of Pisa, Italy

The Leaning Tower of Pisa in Italy is one of the world's most famous buildings. It leans because it was built on soft ground. Soon after construction began in 1173, it started sinking, but it sank more on one side.

By 1990 the tower was about to collapse. Heavy lead bars were piled up on one side to press down the ground and stop the tower from leaning more. In 1999 holes were drilled into the ground on one side of the tower and soil was removed. The holes made the tower sink on one side, tilting it back and saving it from collapse.

The famous Leaning Tower of Pisa leans because it has sunk into the ground unevenly.

What happens to soil during an earthquake?

An earthquake can crack the ground open and bring buildings tumbling down. It can also affect soil in a very strange way. When soil containing a lot of sand and water is shaken by an earthquake, it can behave like a liquid. It flows like water. Heavy things on top of the soil, including buildings, can sink into it. This is called liquefaction.

What are soil's hidden dangers?

Dangers can be hidden underneath soil. In places where there was **mining** in the past, forgotten mine shafts, or tunnels, can collapse without warning and leave a hole in a road or in someone's lawn. In places where wars have been fought, land mines, which are bombs hidden in the soil, still remain. Many people are injured and crippled by stepping on land mines laid during recent wars in eastern Europe and Africa. Bombs that were dropped during World War II are also still found from time to time.

In a landslide, tons of soil can fall down a hillside.

27

CASE STUDY:
The American Dust Bowl

The lives of thousands of people in the United States were affected by a change in the soil. In the 1930s, a huge area of land in the United States was plowed in order to grow wheat on it. Unfortunately, soon after the land was plowed, the worst drought in history began. Without grass, roots, **organic** matter, or water to hold the soil together, it quickly dried out. Strong winds began blowing the soil away. The area of land became known as the Dust Bowl. Almost 865 million tons of soil were blown away in 1935 alone.

The soil became so poor that nothing would grow in it. Farmers were unable to make a living from the land and thousands left to look for work elsewhere. Rain finally returned and ended the drought in 1939.

A drought in the 1930s changed much of the farmland in the United States to dust.

28

Will Soil Ever Run Out?

Soil will not run out because new soil is constantly being made and **recycled** in nature. Although soil will not run out, it is not always in the place where it is needed most. Wind and water can move it away from farmland.

How can we look after soil?

Soil has to be in good condition with plenty of nutrients and **organic** matter to grow healthy crops, so we need to take care of soil that is used to produce food. Organic food production keeps soil in the most natural condition. **Pollution** can damage soil, so we must also protect the land from pollution.

Using organic farming methods is one way of protecting soil from pollution. For example, ladybugs can eat unwanted pests instead of killing the pests with chemicals.

Glossary

erosion slow removal or destruction of soil by the action of wind or water

fertile producing much vegetation or crops

fertilizer substance that is used to make soil more fertile

fungus type of living thing including mushrooms and toadstools that feeds on plants or animals. More than one fungus is called fungi.

glacier body of ice that slowly moves down a valley or slope or across a land surface

habitat natural home of a plant or animal

humus dead and partly rotted parts of plants or animal matter found in soil

inlet small or narrow bay

limestone chalky sedimentary rock formed from animal remains

litter zone surface of soil, covered with dead leaves and plant parts

microbe tiny living organism, that can only be seen by using a microscope

mining digging into the ground to reach valuable materials such as coal

organic of, related to, or obtained from living things. All the material in soil that is alive or was once alive is organic. The sand, clay, and silt in soil are not organic, because they are nonliving.

organism living plant or animal

particle very small bit of something

pollution harmful or poisonous substances in nature, usually produced by the activities of humans

process to change a material by a series of actions or treatments or the method by which a material is changed

recycle to process for reuse instead of using materials only once and then throwing them away

reservoir place where something, such as water, is kept for use in the future

sediment material, such as silt, that settles to the bottom of water

silt very fine particles left as sediment in water. Silt is found in rivers and the sea.

volcano vent, or hole, in Earth's crust from which melted or hot rock and steam come out

More Books to Read

Ballard, Carol. *Soil.* Chicago: Raintree, 2004.

Bocknek, Jonathan. *The Science of Soil.* Milwaukee, Wis.: Gareth Stevens, 1999.

Ditchfield, Christin. *Soil.* Danbury, Conn.: Scholastic Library, 2003.

Farndon, John. *Life in the Soil.* Farmington Hills, Mich.: Gale Group, 2004.

Grolier Educational Staff. *Rocks, Minerals, and Soil.* Danbury, Conn.: Scholastic Library, 2001.

Morris, Neil. *Rocks and Soils.* North Mankato, Minn.: Chrysalis Education, 2002.

Silverstein, Alvin. *Life in a Bucket of Soil.* Minneapolis, Minn.: Sagebrush Education, 2000.

Stewart, Melissa. *Soil.* Chicago: Heinemann Library, 2002.

Index